IT'S TIME TO LEARN ABOUT STOICHIOMETRY

It's Time to Learn about Stoichiometry

Walter the Educator

Silent King Books
A WhichHead Entertainment Imprint

Copyright © 2024 by Walter the Educator

All rights reserved. No part of this book may be reproduced in any manner whatsoever without written permission except in the case of brief quotations embodied in critical articles and reviews.

First Printing, 2024

Disclaimer

This book is a literary work; the information is not about specific persons, locations, situations, and/or circumstances unless mentioned in a historical and factual context. Any resemblance to real persons, locations, situations, and/or circumstances is coincidental. This book is for entertainment and informational purposes only. The author and publisher offer this information without warranties expressed or implied. No matter the grounds, neither the author nor the publisher will be accountable for any losses, injuries, or other damages caused by the reader's use of this book. The use of this book acknowledges an understanding and acceptance of this disclaimer.

It's Time to Learn about Stoichiometry is a collectible little learning book by Walter the Educator that belongs to the Little Learning Books Series. Collect them all and more books at WaltertheEducator.com

STOICHIOMETRY

INTRO

Stoichiometry is a cornerstone of chemistry, providing the tools to quantify and predict the outcomes of chemical reactions. By understanding stoichiometry, chemists can determine the exact proportions of elements and compounds involved in chemical processes. This understanding is essential not only in laboratory settings but also in industrial applications where precise measurements are critical. This little book explores the concept of stoichiometry, its underlying principles, and its practical applications, offering a comprehensive guide to this fundamental aspect of chemistry.

It's Time to Learn about Stoichiometry

Understanding the Basics of Stoichiometry

At its core, stoichiometry deals with the quantitative relationships between reactants and products in a chemical reaction. The term "stoichiometry" itself derives from the Greek words "stoicheion," meaning element, and "metron," meaning measure. This etymology reflects the essence of stoichiometry: it is the measurement of elements involved in chemical reactions.

A chemical equation represents a chemical reaction, where the reactants are transformed into products.

It's Time to Learn about
Stoichiometry

For example, the combustion of methane in oxygen can be represented by the equation:

$$CH_4 + 2O_2 \rightarrow CO_2 + 2H_2O$$

In this equation, one molecule of methane (CH_4) reacts with two molecules of oxygen (O_2) to produce one molecule of carbon dioxide (CO_2) and two molecules of water (H_2O). The coefficients (numbers before the chemical formulas) represent the molar ratio of the substances involved in the reaction.

It's Time to Learn about
Stoichiometry

The Mole Concept

Central to stoichiometry is the mole concept. A mole is a unit used to quantify the amount of a substance, defined as the amount containing the same number of entities (atoms, molecules, ions, etc.) as there are atoms in 12 grams of carbon-12. This number, known as Avogadro's number, is approximately 6.022×10^{23}.

It's Time to Learn about
Stoichiometry

The mole allows chemists to convert between the mass of a substance and the number of particles it contains. For example, the molar mass of water (H_2O) is approximately 18 grams per mole, meaning that one mole of water weighs 18 grams and contains 6.022×10^{23} water molecules.

It's Time to Learn about
Stoichiometry

Balancing Chemical Equations

A crucial step in stoichiometry is balancing chemical equations. A balanced equation ensures that the number of atoms of each element is the same on both sides of the reaction. This is essential because, according to the law of conservation of mass, matter cannot be created or destroyed in a chemical reaction.

It's Time to Learn about
Stoichiometry

To balance an equation, one must adjust the coefficients (the numbers before the chemical formulas) so that the number of atoms for each element is equal on both sides. For instance, consider the combustion of propane (C_3H_8) in oxygen:

$$C_3H_8 + O_2 \rightarrow CO_2 + H_2O$$

To balance this equation, the following steps are taken:

1. Balance the carbon atoms: There are 3 carbon atoms in propane, so we place a coefficient of 3 before CO_2.

$$C_3H_8 + O_2 \rightarrow 3CO_2 + H_2O$$

2. Balance the hydrogen atoms: There are 8 hydrogen atoms in propane, so we place a coefficient of 4 before H_2O.

$$C_3H_8 + O_2 \rightarrow 3CO_2 + 4H_2O$$

3. Balance the oxygen atoms: There are a total of 10 oxygen atoms on the right side (3 from CO_2 and 4 from H_2O, so we place a coefficient of 5 before O_2.

$$C3H8 + 5O2 \rightarrow 3CO2 + 4H2O$$

The equation is now balanced, with 3 carbon atoms, 8 hydrogen atoms, and 10 oxygen atoms on both sides.

Mole-to-Mole Conversions

Once a chemical equation is balanced, stoichiometry can be used to perform mole-to-mole conversions. These conversions are based on the coefficients in the balanced equation, which indicate the ratios in which the substances react.

It's Time to Learn about
Stoichiometry

For example, consider the reaction between hydrogen (H_2) and oxygen (O_2) to form water:

$$2H_2 + O_2 \rightarrow 2H_2O$$

The balanced equation shows that 2 moles of hydrogen react with 1 mole of oxygen to produce 2 moles of water. Therefore, if we have 4 moles of hydrogen, we can use stoichiometry to calculate the amount of oxygen required:

Amount of O_2 = 4 moles H_2 / 2 moles H_2 × 1 mole O_2 = 2 moles O_2

Similarly, we can calculate the amount of water produced:

Amount of H_2 = 4 moles H_2 / 2 moles H_2 × 2 moles H_2O = 4 moles O_2

Mass-to-Mass Conversions

Stoichiometry also enables mass-to-mass conversions, allowing chemists to calculate the mass of reactants and products involved in a reaction. This involves converting the mass of a substance to moles, performing the mole-to-mole conversion, and then converting the moles back to mass.

It's Time to Learn about
Stoichiometry

For instance, consider the reaction of calcium carbonate $CaCO_3$ decomposing into calcium oxide (CaO) and carbon dioxide (CO_2):

$$CaCO_3 \rightarrow CaO + CO_2$$

To determine how much calcium oxide is produced from 100 grams of calcium carbonate, we follow these steps:

1. **Convert mass to moles**: First, calculate the molar mass of CaCO3) (40.08 for Ca, 12.01 for C, and 3 × 16.00 for O):

Molar mass of $CaCO_3 = 40.08 + 12.01 + 48.00 = 100.09$ g/mol

Now, convert the mass to moles:

Moles of $CaCO_3$ = 100 g / 100.09 g/mol ≈ 1.00 mole

2. **Mole-to-mole conversion**: According to the balanced equation, 1 mole of $CaCO_3$ produces 1 mole of CaO.
3. **Convert moles to mass**: Finally, calculate the mass of 1 mole of CaO (40.08 for Ca and 16.00 for O):

Molar mass of CaO = 40.08 + 16.00 = 56.08 g/mol

Thus, 100 grams of calcium carbonate will produce approximately 56.08 grams of calcium oxide.

Limiting Reactants and Excess Reactants

In many chemical reactions, the reactants are not always present in the exact proportions required by the balanced equation. One reactant may be completely consumed before the others, limiting the amount of product that can be formed. This reactant is known as the limiting reactant, while the other reactants are present in excess.

It's Time to Learn about
Stoichiometry

To determine the limiting reactant, follow these steps:

1. **Calculate the moles of each reactant**: Convert the mass of each reactant to moles using their molar masses.
2. **Use the balanced equation to determine the mole ratio**: Compare the mole ratio of the reactants with the ratios required by the balanced equation.
3. **Identify the limiting reactant**: The reactant that produces the smallest amount of product is the limiting reactant.

It's Time to Learn about
Stoichiometry

For example, consider the reaction between nitrogen (N2) and hydrogen (H2) to form ammonia (NH3):

$$N_2 + 3H_2 \rightarrow 2NH_3$$

Suppose we have 28 grams of nitrogen and 10 grams of hydrogen. To find the limiting reactant:

1. **Convert the masses to moles**:
   - Moles of N_2: 28 g / 28.02 g/mol ≈ 1.00 mole
   - Moles of H_2: 10 g / 2.02 g/mol ≈ 4.95 moles
2. **Determine the mole ratio**: According to the balanced equation, 1 mole of N_2 reacts with 3 moles of H_2. However, we have more than enough hydrogen, meaning nitrogen is the limiting reactant.
3. **Calculate the amount of product**: Using the mole ratio from the balanced equation, 1 mole of N_2 will produce 2 moles of NH_3.

Yield: Theoretical, Actual, and Percent Yield

In practice, the amount of product obtained from a chemical reaction is often less than the theoretical yield, which is the maximum amount of product predicted by stoichiometry. The actual yield is the amount of product actually obtained from the reaction, and the percent yield is a measure of the efficiency of the reaction:

Percent Yield= Actual Yield / Theoretical Yield ×100%

It's Time to Learn about
Stoichiometry

For example, if the theoretical yield of a reaction is 50 grams, but the actual yield is only 45 grams, the percent yield would be:

Percent Yield = 45 g / 50 g × 100% = 90%

It's Time to Learn about
Stoichiometry

Applications of Stoichiometry

Stoichiometry is not just a theoretical concept; it has practical applications in various fields:

It's Time to Learn about Stoichiometry

Chemical Manufacturing: In industries, stoichiometry is used to optimize the production of chemicals. By calculating the exact amounts of reactants needed, manufacturers can minimize waste and reduce costs.

It's Time to Learn about
Stoichiometry

Pharmaceuticals: In drug manufacturing, stoichiometry ensures that the correct proportions of ingredients are used to produce medications with the desired efficacy and safety.

It's Time to Learn about
Stoichiometry

Environmental Science: Stoichiometry is applied in environmental studies to quantify pollutant emissions and understand chemical processes in the atmosphere, oceans, and soil.

It's Time to Learn about
Stoichiometry

Biochemistry: Stoichiometry is used to study metabolic pathways, where the precise amounts of reactants and products are crucial for understanding cellular processes.

It's Time to Learn about Stoichiometry

Extrapolating Stoichiometry to Everyday Life

Stoichiometry, at its core, is about understanding the quantitative relationships between different components in a system—essentially, it's about balance and proportions.

This principle can be extrapolated to everyday life situations in various ways. Let's consider the example of cooking, where the idea of "recipe stoichiometry" comes into play.

It's Time to Learn about
Stoichiometry

Cooking and Recipe Stoichiometry

Imagine you're baking a cake. The recipe calls for specific amounts of flour, sugar, eggs, butter, and other ingredients. The success of the cake depends on using the right proportions, just as a chemical reaction depends on the correct ratios of reactants

It's Time to Learn about
Stoichiometry

The Recipe as a Balanced Equation:

In stoichiometry, a balanced chemical equation tells you how much of each reactant is needed to produce a certain amount of product. Similarly, a recipe provides a list of ingredients in specific quantities needed to create a dish.

It's Time to Learn about
Stoichiometry

For example, if a cake recipe requires 2 cups of flour, 1 cup of sugar, 4 eggs, and 1 cup of butter, these amounts are analogous to the coefficients in a chemical equation, representing the proportions required to make the cake (the product).

It's Time to Learn about
Stoichiometry

Scaling Up or Down:

Just as stoichiometry allows you to scale a chemical reaction up or down, you can do the same with a recipe. If you want to make a double batch of cake, you simply multiply each ingredient by 2, ensuring that the ratios remain consistent. Conversely, if you want to make only half a batch, you divide each ingredient by 2.

It's Time to Learn about
Stoichiometry

This concept is similar to how stoichiometry enables chemists to calculate the amounts of reactants needed for different scales of a reaction.

It's Time to Learn about
Stoichiometry

Limiting Ingredients:

In stoichiometry, the limiting reactant is the substance that is completely consumed first, limiting the amount of product that can be formed. In cooking, if you have limited quantities of certain ingredients, they become the limiting factor in how much of the final dish you can make.

It's Time to Learn about
Stoichiometry

For instance, if you only have 2 eggs but the recipe requires 4, the eggs are the limiting ingredient. You'll need to adjust the quantities of all other ingredients accordingly to maintain the proper ratios, just as you would adjust the quantities of other reactants in a chemical reaction.

It's Time to Learn about
Stoichiometry

Yield and Efficiency:

In chemistry, the theoretical yield is the maximum amount of product that could be formed from the given reactants, while the actual yield is what you actually obtain. The percent yield measures the efficiency of the reaction.

It's Time to Learn about
Stoichiometry

Similarly, when cooking, the recipe provides a theoretical yield—how much food you expect to produce. However, the actual yield might be less due to spillage, ingredients sticking to the bowl, or other factors. The concept of percent yield can be applied here as well, where you compare the amount of food you actually made to what the recipe promised.

It's Time to Learn about
Stoichiometry

Conservation of Mass:

The principle of conservation of mass states that matter cannot be created or destroyed in a chemical reaction. In cooking, this is reflected in the idea that the total amount of ingredients you start with will equal the total amount of food produced (plus any waste or loss).

It's Time to Learn about
Stoichiometry

For example, if you're making a stew, all the vegetables, meat, and broth you put in the pot remain in the stew or are transformed in a way that nothing is lost, just as the total mass of reactants equals the total mass of products in a chemical reaction.

It's Time to Learn about
Stoichiometry

Stoichiometry, though a chemical principle, finds its analogs in everyday life, particularly in situations where proportions, balance, and the transformation of materials are involved.

It's Time to Learn about
Stoichiometry

Cooking is a perfect example where the ideas of balanced equations, limiting ingredients, scaling recipes, and yield can be easily understood and applied.

It's Time to Learn about
Stoichiometry

By recognizing these parallels, one can appreciate how the principles of chemistry extend beyond the laboratory and into the kitchen and everyday activities.

It's Time to Learn about
Stoichiometry

Stoichiometry is a vital concept in chemistry, providing the quantitative framework needed to understand and predict the outcomes of chemical reactions.

It's Time to Learn about
Stoichiometry

By mastering stoichiometry, chemists can calculate the amounts of reactants and products, determine limiting reactants, and assess the efficiency of reactions.

It's Time to Learn about
Stoichiometry

This understanding is essential not only in academic research but also in various industrial and environmental applications.

It's Time to Learn about
Stoichiometry

Stoichiometry, therefore, is not just a tool for chemists but a critical component of many scientific and engineering disciplines.

It's Time to Learn about
Stoichiometry

ABOUT THE CREATOR

Walter the Educator is one of the pseudonyms for Walter Anderson. Formally educated in Chemistry, Business, and Education, he is an educator, an author, a diverse entrepreneur, and he is the son of a disabled war veteran. "Walter the Educator" shares his time between educating and creating. He holds interests and owns several creative projects that entertain, enlighten, enhance, and educate, hoping to inspire and motivate you. Follow, find new works, and stay up to date with Walter the Educator™

at WaltertheEducator.com

www.ingramcontent.com/pod-product-compliance
Lightning Source LLC
LaVergne TN
LVHW051925060526
838201LV00062B/4693